**SUN GUN TELESCOPE for Solar Astronomy
By Bruce Hegerberg**

Contents

Preface — 3

How To Build A Sun Gun and Variations of It

Big Easy Sun Gun — 4

Rear Projection Assembly — 5

Sun Gun (Original Design) — 11

Sun of a Gun — 17

Sun Shots Gallery — 22

Other Astronomy Related Projects

Daystar Quark — 32

Lunt Solar Scope Plus — 34

Lunt with Moonlite Focuser — 37

Lunt Ipad 2 — 38

Coronado Maxscope 40 — 39

Stanford Radio Telescope — 42

Solar Radio Telescope — 43

Solar Observatory — 54

Gallery of Sun Guns

Donations to Museums Across the US — 55

Nature Causes

The Environment — 63

Preface

This book is a collection of how to instructions and links related to the Sun Gun and other astronomy experiments from my father's website www.sunguntelescope.com. The words are entirely his - all from his website, with just a few comments from me from time to time.

My father dedicated his life to the study of astronomy and in particular to solar studies. His greatest wish was to inspire children to explore astronomy. It was that passion for education that led him to inventing the Sun Gun and sharing it with the world. It is my hope that in publishing this book that his legacy will live on well after I am gone.

I know that in time many of the links contained in this book may well become obsolete, but it is my hope that the ideas contained herein will provide inspiration for generations to come to both replicate and advance upon the concepts presented, given whatever access you have to materials wherever and whenever you choose to begin your path of exploration.

My father, Bruce Hegerberg, was from Jackson, Michigan. Born in the late 1940s he was the son of a prison guard and a store clerk. He was inspired by science at an early age and was entirely self-taught. After graduating high school, like most of the boys in his neighborhood, college was a expense far beyond their financial means, so he entered the US Marine Corps. My father was a proud Marine and often told us stories of his days in the Corps. To this day we donate whenever we can to the Marine Corps Scholarship Foundation in his memory and would ask that you do the same if you would like to honor him too. He later worked as an Industrial Controls Consultant, again teaching himself computer languages and electronics, to start his own business.

Up until his last days he was tinkering with whatever was laying around the house, coming up with new ideas all the time of how to get kids interested in science and how to make astronomy exploration affordable for adults.

My father passed away from cancer in 2015. He was the light of our lives, and it is of such great pride to my mother and I to see how children and adults alike at museums across the United States continue to be inspired by his invention. As my father always said, Enjoy!

Big Easy Sun Gun

Parts List

Tripod
http://www.amazon.com/gp/product/B000E0PPG0/ref=oh_details_o00_s00_i00
Tripod Adapter
http://www.amazon.com/gp/product/B0007SL856/ref=oh_details_o00_s01_i00
Rear Projection Assembly
Explained on next page
Finder
http://www.farlaboratories.com/dyna-hp1.html

I use a Celestron Nexstar 102GT telescope for the Big Easy Sun Gun. You can find the telescope, with mount on Ebay . Remove the dovetail bar from the telescope and remount it next to the focuser, using 1/8" mounting holes, use this bar for the scope. I found another dovetail bar on Ebay and mounted this next to the focuser. The only part you have to build is the rear projection assembly, which is not too difficult. Enjoy!

Rear Projection Assembly

Parts List

Urn
http://www.amazon.com/Allied-Precision-Industries-CRU23BLK-Renaissance/dp/B007IHIZ32/ref=sr_1_97?s=lawn-garden&ie=UTF8&qid=1335965714&sr=1-97

Plate
https://www.daystarfilters.com/AdapterPlates.shtml

Screen
http://www.rosebrand.com/product703/Projection-Screen-and-Rear-Projection-Screen.aspx?cid=218&idx=1570&tid=1&info=Screen%2bby%2bthe%2bYard

Filter Holder
http://www.adorama.com/SZ25216.html

Mirror Clips
http://www.homedepot.com/h_d1/N-5yc1v/R-202210185/h_d2/ProductDisplay?catalogId=10053&langId=-1&keyword=mirror+clip&storeId=10051

Hoop
http://www.ebay.com/itm/Wood-EMBROIDERY-HOOP-23-X-3-4-/140755634957?pt=LH_DefaultDomain_0&hash=item20c5b08b0d#ht_720wt_92

2 (* Mount screws with washer on head side and washer, lock washer and nut on other.)

Instructions

1. Mount a 2' by 2' piece of rear projection screen in the hoop, dull side up, make it as tight as possible

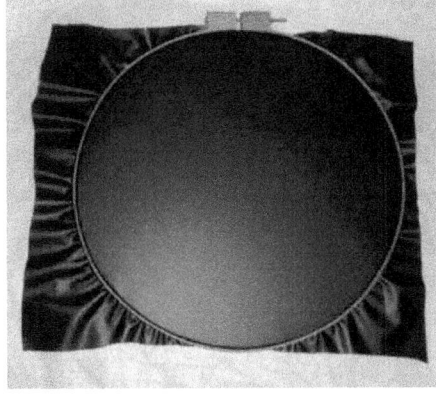
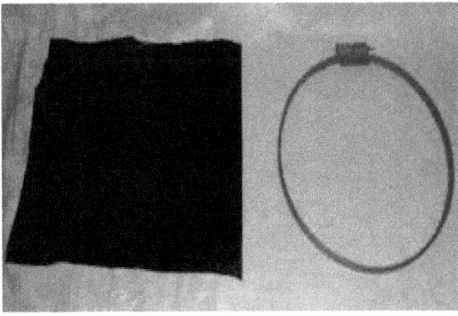

2. Take apart the filter holder and use the bottom ring with four holes. Prime and paint flat black one side, let dry. On the non painted side of the holder coat the outer half with super glue. Press the holder super glue side, straight down into the screen on the hoop, do not move around. Let this set over night.

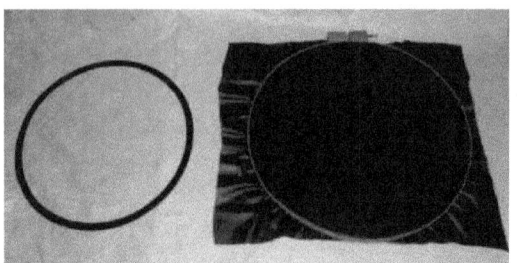 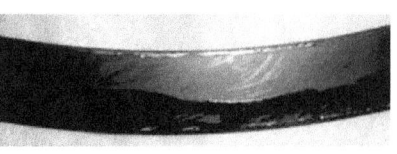

3. When this has set overnight cut out along the filter holder outer edge, with a single edge razor blade.

This is what it will look like when done.

4. Make a bend in one leg of the four mirror clips, Mount the clips in the four holes on the filter holder, mount on the under non painted screen side, use *6-32 1/2 screws. Set this aside for now.

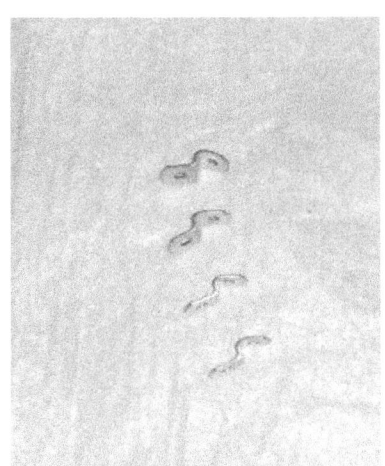
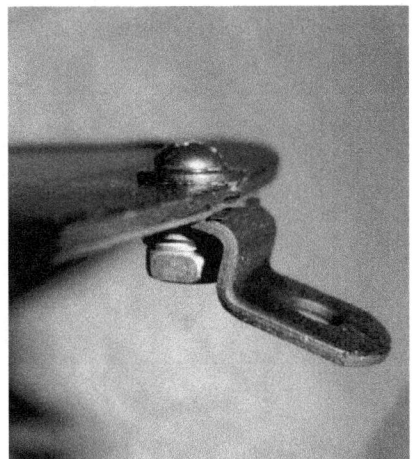

Rear Projection Assembly Urn

1. Turn urn upside down. drill a pilot hole in the center of the bottom. With a 1.75" hole saw, cut a hole in the bottom.

2. Center the plate Q3RFCTM over the 1.75" hole. Mark the three holes, drill with 5/32 bit. Mount with three #8-32 X 5/8 screws. Get two 3/8" plastic plugs, install in the two drainage holes in the urn bottom, may have to be trimmed to fit.

3. Turn urn upright, center screen on top. Mark holes, and drill with 5/32 bit. Mount screen with #6-32 X 3/8 screws.

4. I have been using a 10mm 70 degree eyepiece, and a #12 yellow filter, with a 102mm telescope. Mount the eyepiece in a 2" to 1.25 adapter, with t-thread. The eyepiece is mounted sticking out of the adapter, so you can reach focus. This will screw into the adapter on the bottom of the urn. YOU'RE DONE.

Rear Projection Screen Retail Examples

Cousin's
http://store.cousinsvideo.com/41468.html

Rose Brand
http://www.rosebrand.com/product703/Projection-Screen-and-Rear-Projection-Screen.aspx?tid=2&info=REAR%2bPROJECTION

Sun Gun (Original Design)

I wanted a solar scope, safe and portable, for group sun spot viewing. The designs I looked at did not lend themselves to group viewing, or the safety I wanted, especially with children. After many hours at the hardware store I designed the SUN GUN.

The SUN GUN consists of a 60mm dia. 900mm fl. optical tube assembly (Earth and Sky-Telescopes, www.astrosales.com/scopes.html), mounted inside of 3" dia. PVC, connected to a 20" flower planter with a rear projection screen mounted on top. When connecting the PVC parts lightly sand each connection and apply PVC cement to both connections. Home depot PVC was used through out the design, found some suppliers to have a different inside diameter.

Telescope Assembly

Connect a 28.5" piece of 3" PVC to one side of a 3" tee, on the other side connect a 3" cleanout adapter with female threads. Take a 3" cap and connect a 3" piece of 3" PVC to it, this will be a cover for the remaining side of the tee, do not cement it to the tee. Take the optical tube assembly, remove the lens cap, and twist off the light shield. Remove the bracket holding the pinion gear, remove the gear and focusing assembly. Remove the knob that can be unscrewed from the focusing assembly, this will not be used. Obtain two Hoover vacuum cleaner belts B-20, type 48 (Home Depot Part Number 074999020202).

Cut a section out of the belts so they will just snap over the optical Tube Assembly, super glue the belts back together. Install one of the belts on the lens end of the optical tube assembly, then push the assembly through the cleanout adapter, tee, and 28.5" PVC, in that direction. Let the optical tube assembly extended a couple inches beyond the tube, then install the other belt and pull the optical tube assembly back into the PVC until the focuser is centered in the tee, the belts will end up near each end of the optical tube assembly. The optical tube assembly will still be extended about a half inch from the end. Turn the optical tube assembly so that the pinion gear and shilf with one knob can be reinstalled with the bracket. After the shaft with one knob has been installed turn the optical tube assembly so the knob is accessible in the tee.

Install a .965" to 1.25" eyepiece adapter, because the 60mm telescope came with a .965" eyepiece holder. This worked out ok because the eyepiece is extended beyond the 3" cleanout adapter when the rear projection section is removed, giving easy access to the eyepiece. Install a 1" length of 3" PVC inside of a 3" PVC coupling. The light shield from the optical tube assembly is pushed into the 3" PVC coupling, the top end first, until the ridge on the shield meets the 1" length of 3" PVC inside of the coupling. This is installed, but not cemented, on top of the optical tube assembly and the 28.5" PVC giving access to the optical tube assembly for adjustment.

Rear Projection Assembly

Install 3" PVC 3" long pipe into a floor flange, and install a 3" PVC male adapter onto this. On the outside bottom of a 20" flower pot (Home Depot part number 087404560200) center the floor flange and drill four mounting holes. Mount the floor flange on the outside to a closet flange extension ring on the inside. Cut out the bottom of the flower pot using the inside of the closet flange extension ring as a guide. Drill mounting holes in each of the four tabs on a SPEEDTRON 20" photographic light diffuser (SPEEDOTRON part number 25529). Cut out the diffuser that comes with it, the outside ring is only used. Tack the corners of a 2' by 2' Da-Tex rear projection screen (Da-Lite Screen Company 800-622-3737, www.da-lite.com), polished side up, stretched between two nailed down two by fours. Apply super glue to the bottom of the 20" light diffuser ring, press the diffuser ring, glue down, onto the rear projection screen. Hold until set up. Align the diffuser ring with the rear projection screen on the top lip of the flower pot, mark where the mounting holes are in the tabs on the diffuser ring on the flower pot. Remove diffuser ring and drill holes. Set aside, the rear projection screen is the last thing to be mounted in the assembly.

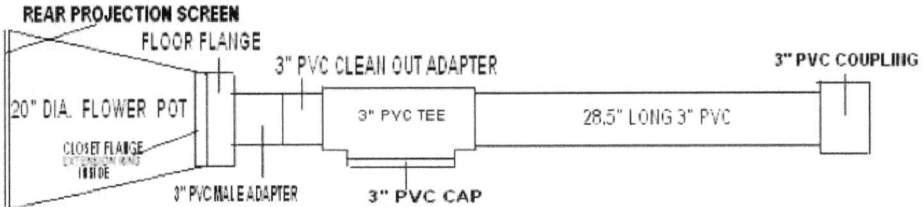

Altazimuth Mount Assembly

On a 3" PVC clean out tee install a 3" PVC 4.5" long piece. Connect a 3" PVC male adapter to this. Install a 3" PVC clean out adapter to a 4" to 3" saddle tee, and screw this onto the 3" PVC male adapter, this being the altitude axis. On the other side of the clean out tee install a 3" PVC 11.5" long piece, and connect to this a 3" to 2" PVC reducer. To the reducer install a 2" to 3/4" PVC threaded reducer. Screw a 3/4" gal nipple 3" long into the threaded reducer. Put a 4 kilos/8.8 lbs bar bell weight on the 3/4" gal nipple, and install a 3/4" gal cap on the nipple. The threaded connection of the 3" PVC clean out tee and the male 3" PVC male adapter on the pedestal make up the azimuth axis.

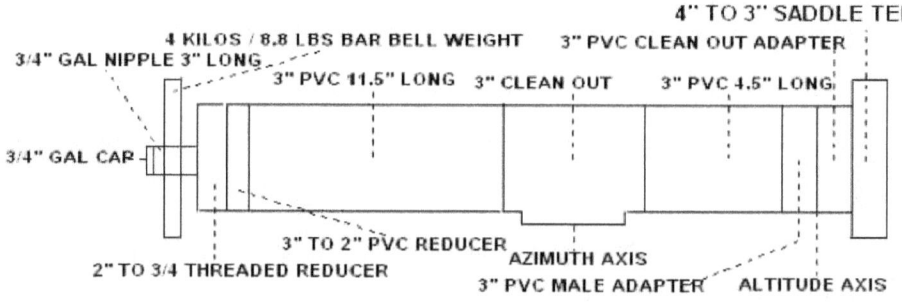

Pedestal Assembly

Mount four 3" casters on the corners of a 2' by 2' by 3/4" piece of plywood. Install a 2" length of 4" PVC on a floor flange, and connect a 4" PVC male adapter to this. Center the floor flange on the 2' by 2' plywood, drill four holes, and secure with nuts and bolts. Cut a ½" piece off the end of a 3" PVC coupling. Install the ½" coupling piece flush to the top, on the inside, of a 48" long piece of 4" PVC. Install a 4" PVC female adapter to the other end of the 48" long piece of 4" PVC. Install the 3" PVC coupling, the ½" piece was cut off from, on to the bottom of a 45" length of 3" PVC. Insert the 45" length of 3" PVC, the end opposite the 3" PVC coupling, through the 4" PVC female adapter on the 48" length of 4" PVC up the inside of the 4" PVC, and out the top through the ½" 3" coupling piece. Sand down the inside of a 4" to 3" rubber flex coupling, otherwise its near impossible to get on. Install the flex coupling over the 4" PVC and the 3" PVC. On the top of the 45" long 3" PVC install a 3" PVC male adapter.

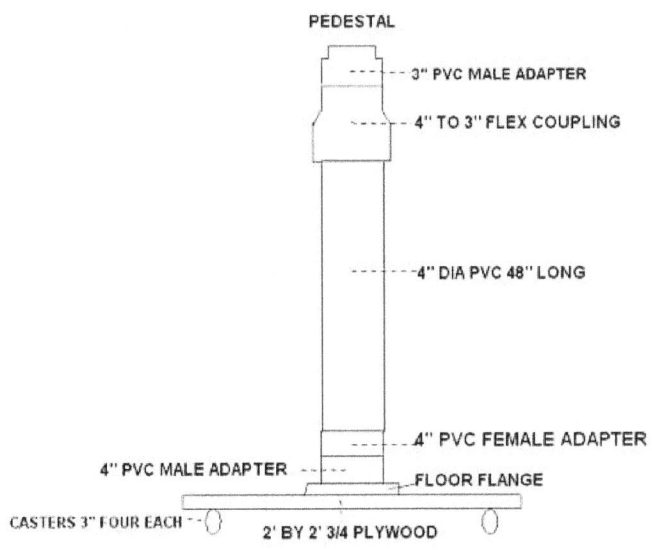

Final Assembly

Mask everything not to be painted. Do not paint the 45" long 3" PVC inside of the pedestal. Prime all the PVC and paint the telescope and projection assembly a bright color, the prototype is florescent yellow. Paint the inside of the flower pot and its rim flat black. Screw the pedestal onto the base. Screw the Altazimuth mount onto the 3" male adapter on the top of the pedestal. Mount the telescope to the altazimuth mount, with the saddle clamps over the 3" tee. Install a 12.5mm eyepiece with a # 12 eyepiece filter in the eyepiece holder. Mount the rear projection screen assembly on top of the flower pot, with the drilled taps on the diffuser, and holes in flower pot top rim. With a 10mm eyepiece and a #15 deep yellow eyepiece filter, a 19" image is obtained with the SUN GUN. Better contrast is obtained with a 12.5mm eyepiece and a #12 yellow eyepiece filter giving a 18" image and a 2" border. Enjoy!

Sun Of A Gun

While sitting on a paint bucket and meditating on the SUN GUN I saw it! The SUN OF A GUN. The SUN OF A GUN is a smaller version of the of the SUN GUN. It will work with any telescope.

Sun Of A Gun Assembly

1. I use a Home Depot Homer's bucket (http://www.homedepot.com/h_d1/N-5yc1v/R-100087613/h_d2/ProductDisplay?catalogId=10053&langId=-1&keyword=homer+bucket&storeId=10051) . Turn the bucket upside down and with a 1.75" hole saw cut a hole in the center of the bottom.

2. Mount a 1.25" clamp connector (http://www.homedepot.com/h_d1/N-5yc1v/R-100192775/h_d2/ProductDisplay?catalogId=10053&langId=-1&keyword=90514&storeId=10051) for nm cable, in this hole with the clamp mounted on the out side. Use two conduit locknuts (http://www.homedepot.com/p/Halex-1-1-4-in-Conduit-Locknuts-2-Pack-96194/100183711#.UjonbBY0FNI), one on the outside the other on the inside. Replace the screws that come with the clamp with #10-24 x 1.50" (http://www.homedepot.com/h_d1/N-5yc1v/R-202706080/h_d2/ProductDisplay?catalogId=10053&langId=-1&keyword=27771&storeId=10051).

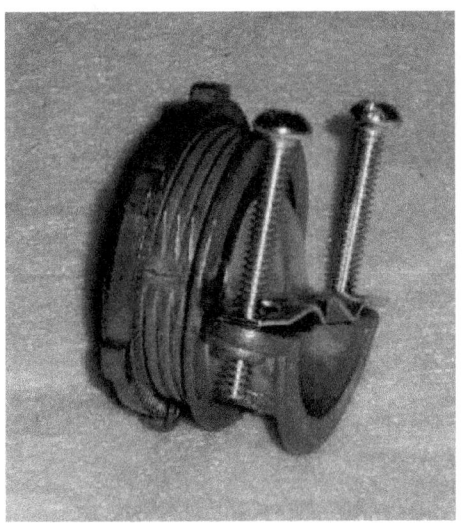 

3. A 15" square piece of rear projection screen is mounted to the top of the paint bucket with a 48" wire tie (http://www.homedepot.com/h_d1/N-5yc1v/R-202905437/h_d2/ProductDisplay?catalogId=10053&langId=-1&keyword=wire+tie+48&storeId=10051) or a large rubber band. Lay the rear projection screen on top of the bucket, dull side facing out. Connect the wire tie so it just fits over the rear projection screen and the top lip of the bucket. Tighten the wire tie under the lip of the bucket and adjust the rear projection screen tight across the top of the bucket, it should look like a drum when done. Cut the excess screen off, leave ½" of screen below the wire tie for future adjustment.

4. Mount a eyepiece to the clamp, 15mm or 18mm with a #12 yellow filter.

5. Use a 36" rubber tarp strap (http://www.homedepot.com/h_d1/N-5yc1v/R-202224213/h_d2/ProductDisplay?catalogId=10053&langId=-1&keyword=rubber+tarp+strap&storeId=10051) through the handle to make the bucket level with the scope. Use on a telescope with a lens of 4" or less, or stop down a larger scope to 4" or less. You may have to have the eyepiece stick out of the focuser one half inch so you can reach focus.

Sun Shots Gallery

Mercury transit of the Sun.

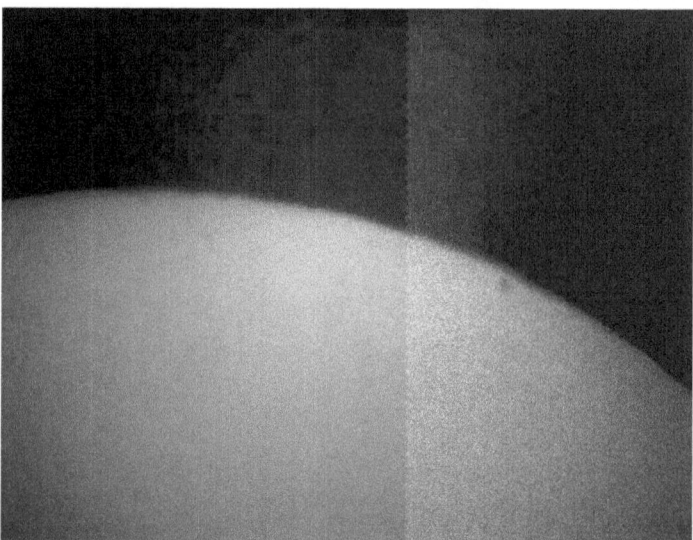

December 25, 2000.

Holographic diffraction grating mounted over a barlow lens.

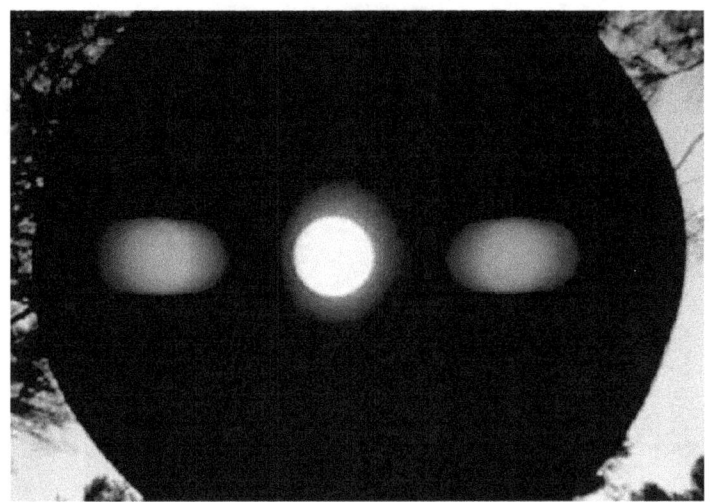

Toowoomba, Australia. Lindsay Ball and Dr. Brad Carter.

Venus June 5, 2012.

From various events in the Atlanta area and at our home.

One of my father's favorite photos. It was taken by the Smithsonian Institute in Washington DC at their observatory. It is a photo of the Sun Gun inside the observatory superimposed upon an external view of the observatory on the National Mall.

My father had a great sense of humor. He asked my mother to move the Sun Gun in the direction of the sun while he adjusted something. She couldn't quite figure out where it was. I captured the moment in a photo and dad put the caption on it.

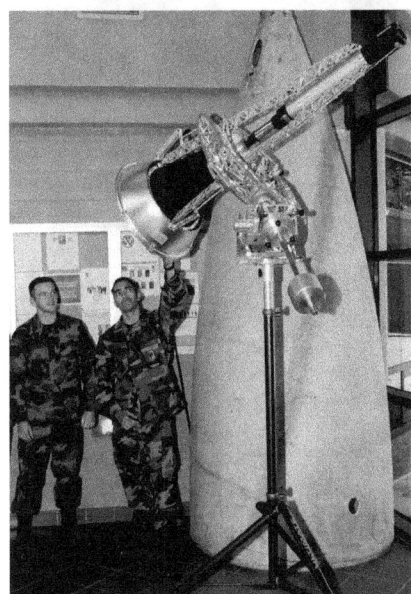

A super fancy Sun Gun built by the military.

Daystar Quark

Quark mounted on a Skywatcher 120mm F5 with a 2" UV/IR cut filter.

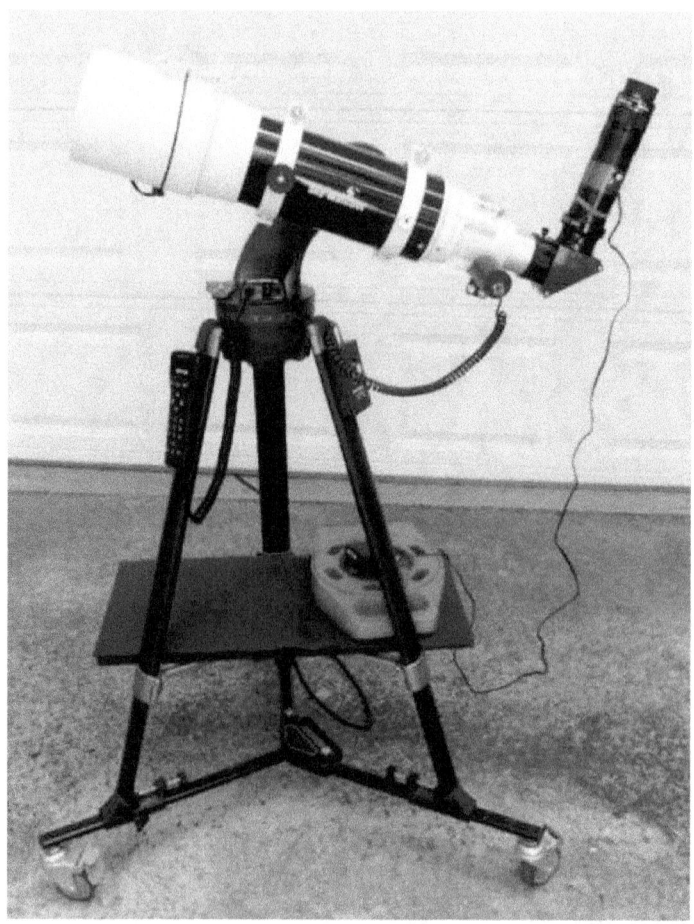

Quark mounted on Astro Tech 66mm F5.

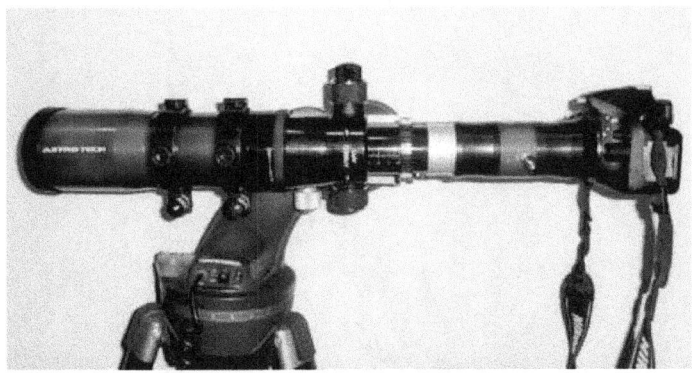

Quark mounted on a Meade ETX 60.

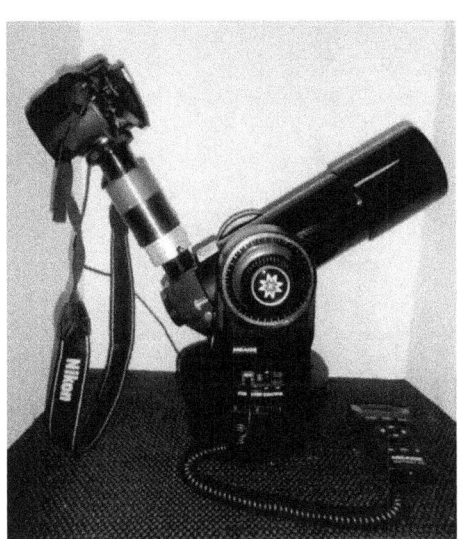

I have only been able to checkout the Skywatcher 120mm setup so far . I am very pleased with the Daystar Quark.

Lunt Solar Scope Plus

I wanted a setup that had everything together, got tired of dragging half the house outside.

The dolly I got from Ebay (http://cgi.ebay.com/Pro-Heavy-Duty-Video-Camera-Photo-Tripod-Dolly-FREE-bag-/330519981794?pt=LH_DefaultDomain_0&hash=item4cf486aee2#ht_3293wt_1 141). It folds up and I like the way the 3" wheels lock and unlock.

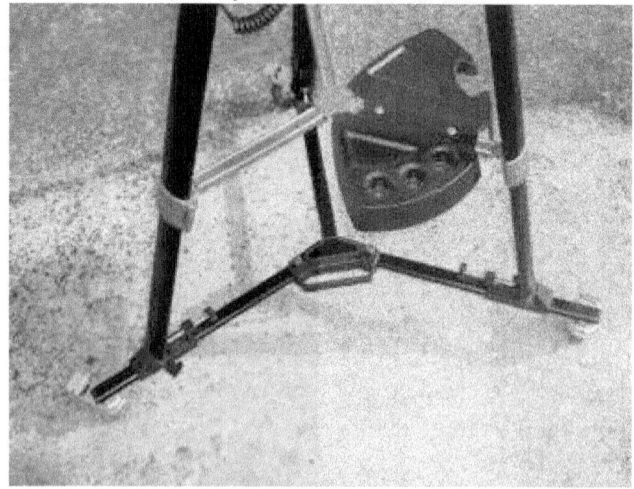

The mount is a Coronado DSM mount
(http://telescopes.net/doc/2500/mftr/Coronado/item/DSM) for solar scopes,
works great for this size scope. I got it from Woodland Hills Camera and
Telescope.

A pair of Orion 90mm rings
(http://www.telescope.com/control/product/~category_id=mount_accessories/~
pcategory=telescope_mounts/~product_id=07370/~sSearchSession=5a1e06c
8-8411-438c-b812-288e3c72c7ea) are used to mount the
computer, Acer Aspire One, to the telescope. A tee plate
(http://www.homedepot.com/h_d1/N-5yc1v/R-
202034097/h_d2/ProductDisplay?storeId=10051&keyword=030699150687&js
pStoreDir=hdus&Nu=P_PARENT_ID&navFlow=3&catalogId=10053&langId=-
1&ddkey=Search)
was mounted to the rings with 1/2" spacers. Velcro was placed on top of this,
to mount the computer too.

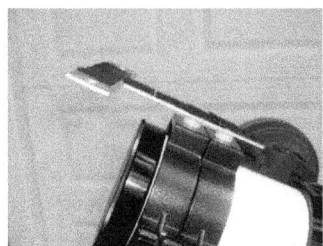  

The Sol Searcher was mounted to the bottom of plate the telescope is mounted on. I found a finger mouse on Ebay, that works well with this setup. With the computer display up it acts as a sun shield. I really like this setup I can be out and running in a couple minutes.

Enjoy.

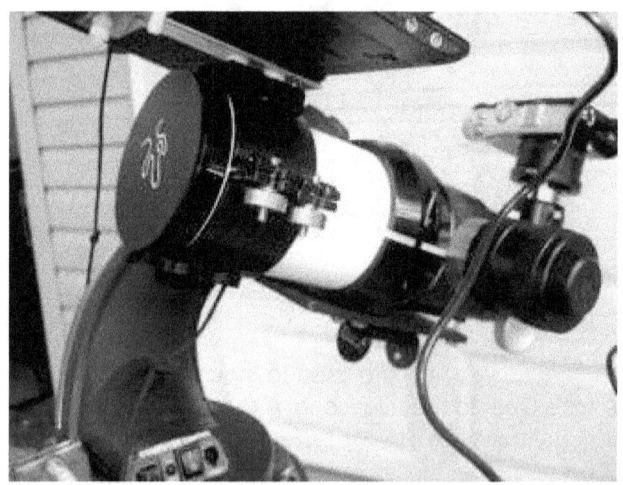

Lunt With Moonlite Focuser

Really looks and works great.

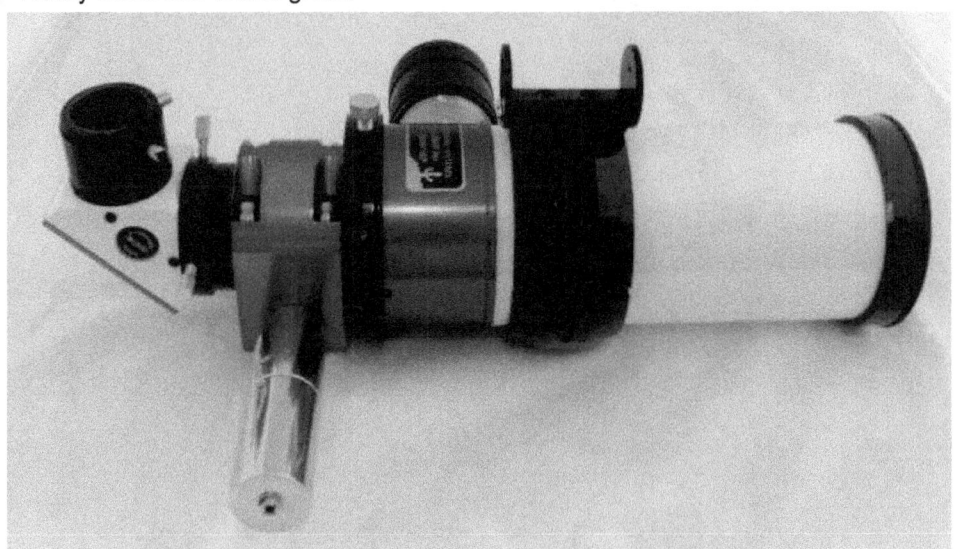

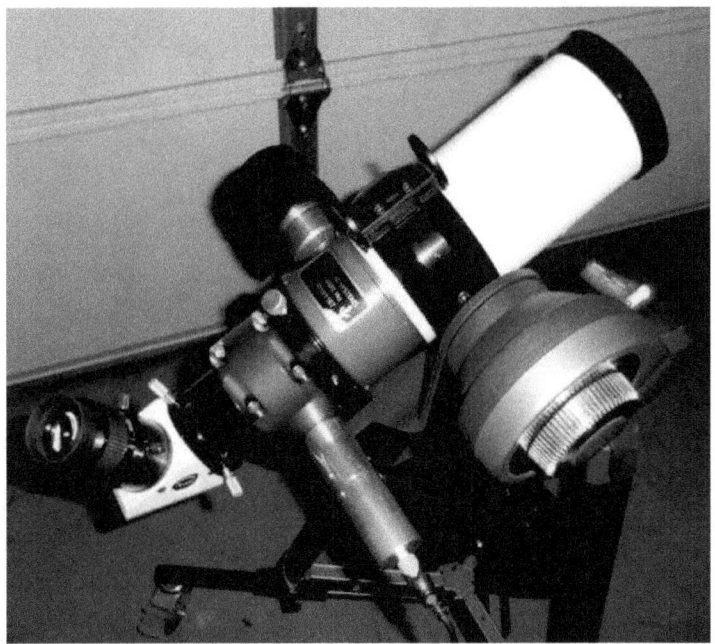

Lunt Ipad 2

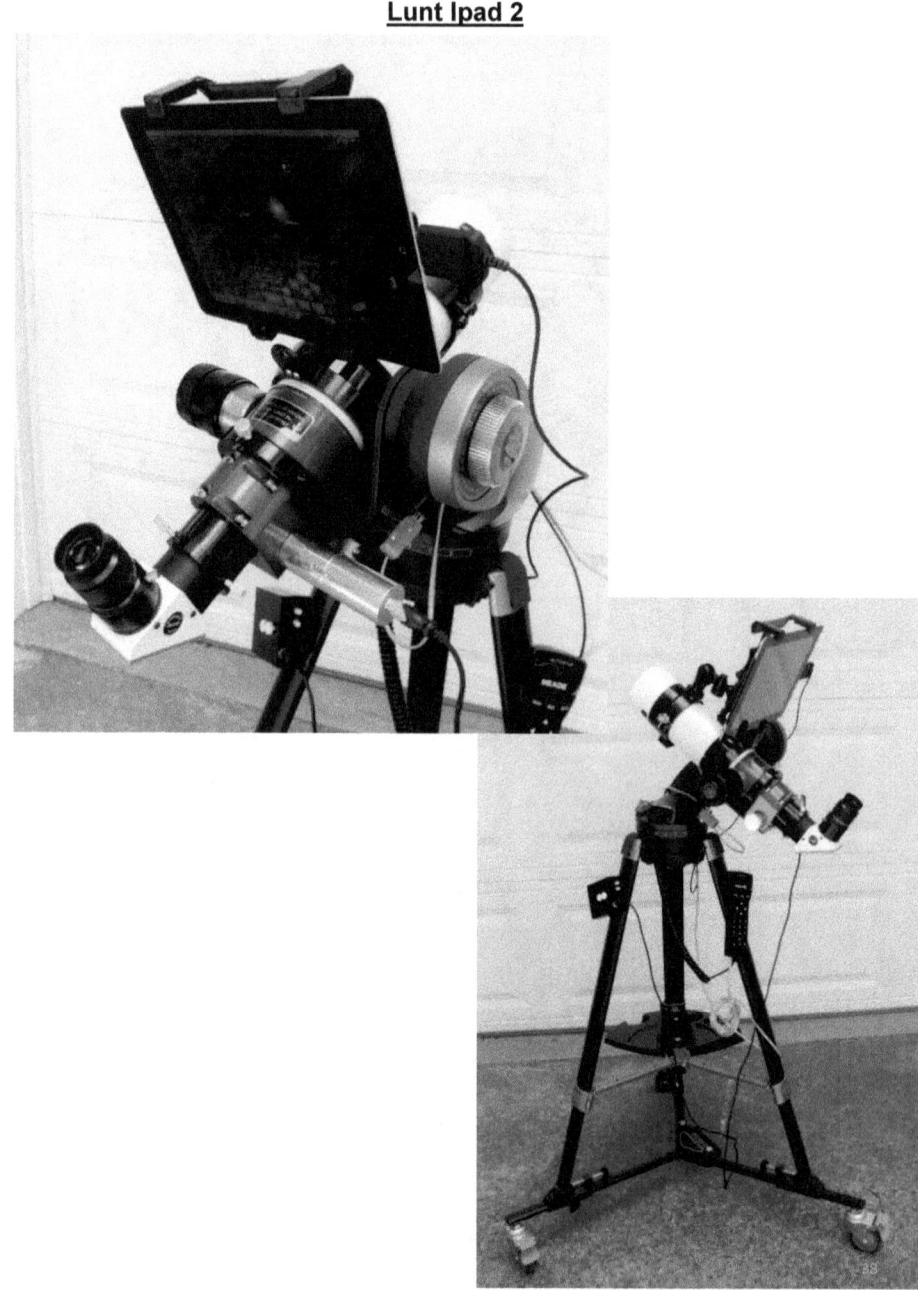

Coronado Maxscope 40

I have enjoyed this telescope more than any scope that I have owned. If you have any interest in solar astronomy, or want to get into it, get one. The only problem that I have had is getting a decent photo. After many photos I decided to use the red channel for the proms and the green channel for the surface, from one over exposed photo, as described below.

 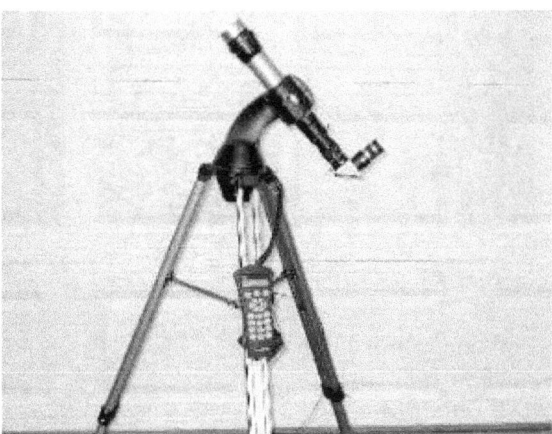

SET UP:

The MaxScope 40 (http://www.coronadofilters.com/) was installed on a Celestron NexStar 60 mount (http://www.celestron.com/main.htm). A Tele Vue Sol-Searcher finder (http://www.televue.com/) was mounted on the clamshell of the Nexstar, this is a must have, as it greatly reduces the time to get the image into the field of view. A Coronado 18mm Cemax eyepiece (http://www.coronadofilters.com/)for visual observing, and a 24mm Cemax eyepiece for photographic use. A Nikon CoolPix 4500 camera, with Scoptronix adapters, (http://www.scopetronix.com/) for taking photos. All of the above can be purchased from Anacortes (http://www.buytelescopes.com/).

THE BIG PICTURE:

Before   After

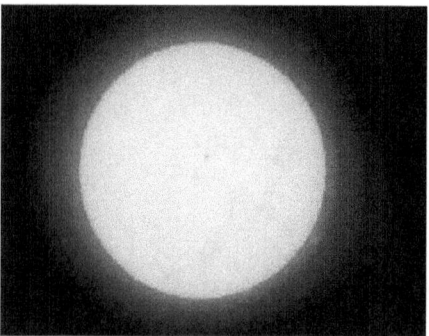 

CAMERA SETTINGS

Nikon CoolPix4500
2003/07/03 13:14
TIFF-RGB(8-bit)
Image Size: 2272 x 1704
Color
Converter lens: none
Focal Length: 28.9
Exposure Mode: Shutter Priority
Metering Mode: Multi-Pattern
1/30 sec- f/9.7
Exposure Comp: 0 EV
Sensitivity: ISO 100
White Balance: Auto
AF Mode: AF-S
Tone Comp: Normal
Flash Sync Mode: Not Attached

COMPUTER PROCESSING OF IMAGE
Paint Shop Pro 7 was used to process this image.

1. File-Open-(select image)-Open
2. Select image
3. Colors-Split Channel-Split to RGB
4. Close Blue background-Do not save
5. Select red Background
6. Colors-Adjust-Brightness/contrast-(set Brightness -50/ Contrast +50)-OK
7. Effects-Sharpen-Unsharp mask-(Radius 30/Strength 100/Clipping 0)-OK
8. Colors-Adjust-Highlight/Midtone/Shadow-(Shadow 50/Midtone 50/Highlight 80)-OK
9. Colors-Combine Channel-Combine from RGB-(Make all Three channels the redsource)-OK
10. Select Green background.
11. Effects-Sharpen-Unsharp Mask-(Radius 5/Strength 200/Clipping 0)-OK
12. Colors-Combine Channel-Combine from RGB-(Make all three channels the greensource)-ok
13. Select the green combined image.
14. On the Green combined image click the magic wand, from the tool bar on the left, on the dark background outside of the solar disk.
15. Selections-Invert
16. Edit-Copy
17. Select the Red combined image.
18. Image-Resize-Click Percentage of original-(width 99/ Height 99)-OK
19. Edit-Paste-As new layer
20. Align the solar disk with the prom background.
21. Layers-Merge-Merge all
22. Colors-Adjust-Levels-(Input Levels, RGB as is, Red as is, Green center slider move to .25, too the right, Blue left and center slider move all the way to the right)-OK

Stanford Radio Telescope

This is my take on the Stanford radio telescope . SID (http://solar-center.stanford.edu/SID/sidmonitor/).

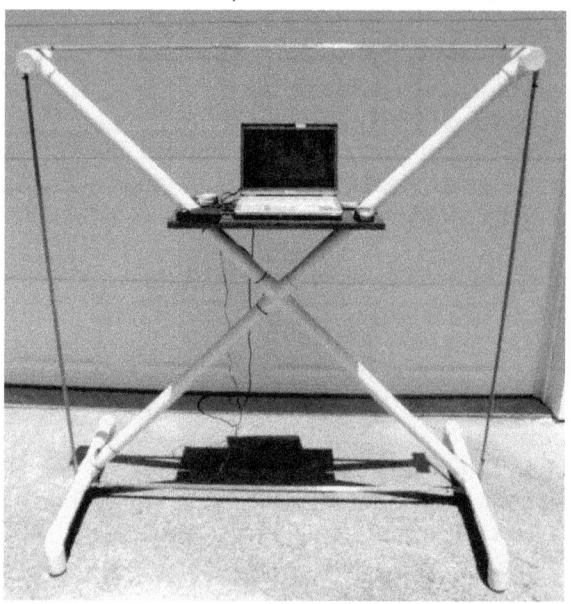

Solar Radio Telescope

I assembled the JOVE receiver, at 20.1 mhz, and the GYRATOR II receiver, at 24 khz. The two receivers outputs were applied to a DATAQ (http://www.dataq.com/products/startkit/di194rs.htm) adc.Using the WinDaq 194 software (http://www.dataq.com/products/software/acquisition.htm) supplied with the adc, the two signals can be compared.

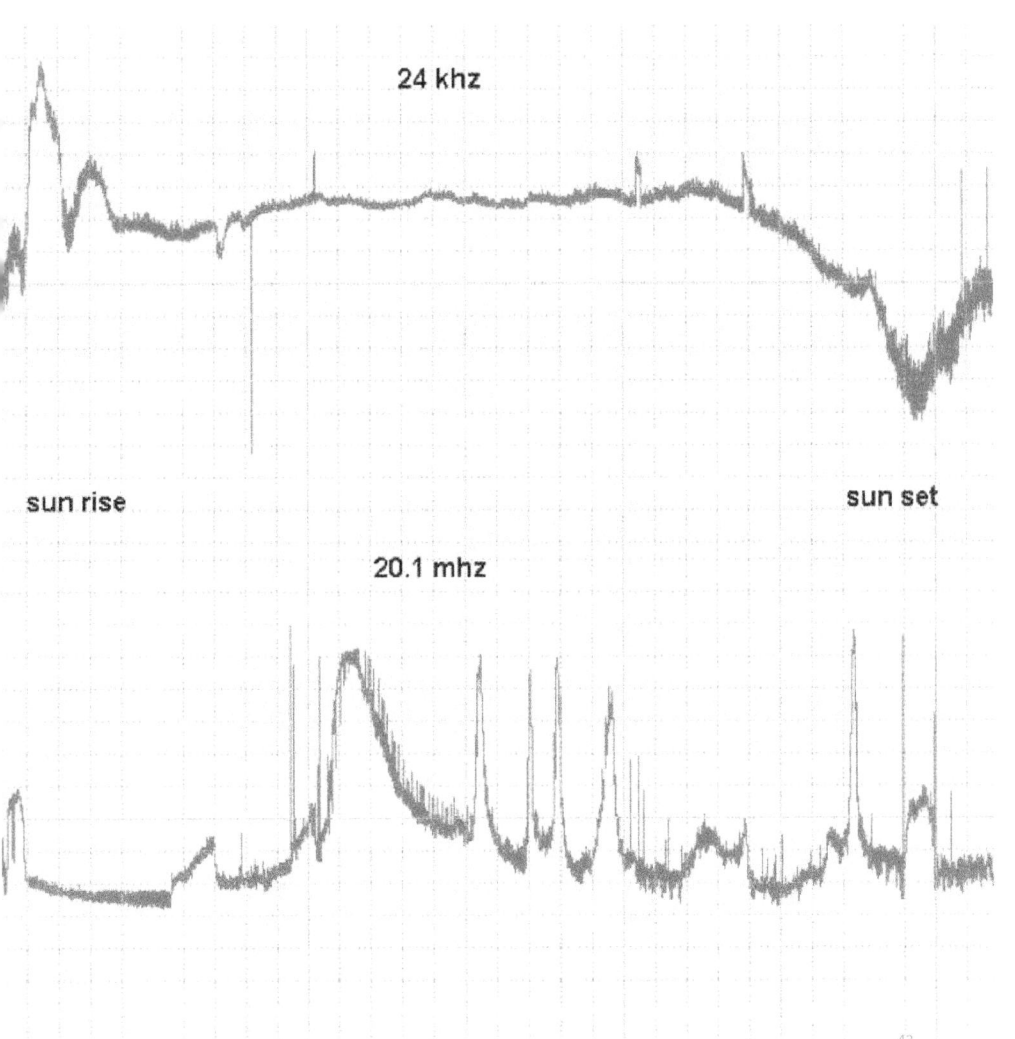

PARTS

1   JOVE 20.1 receiver   JOVE
(http://radiojove.gsfc.nasa.gov/telescope/)

1   GYRATOR 24 KHZ RECEIVER   AAVSO
(http://www.aavso.org/gyrator-iii-vlf-receiver)

1   WHIP ANTENNA   PART NO. 21-903   RADIO SHACK
(http://www.radioshack.com/product/index.jsp?productId=2102428)

1   LOOP ANTENNA   RADIO SHACK
(http://www.radioshack.com/product/index.jsp?productId=2102428)

1   MINI AMPLIFIER   PART NO. 277-1008   RADIO SHACK
(http://www.radioshack.com/product/index.jsp?productId=2062620)

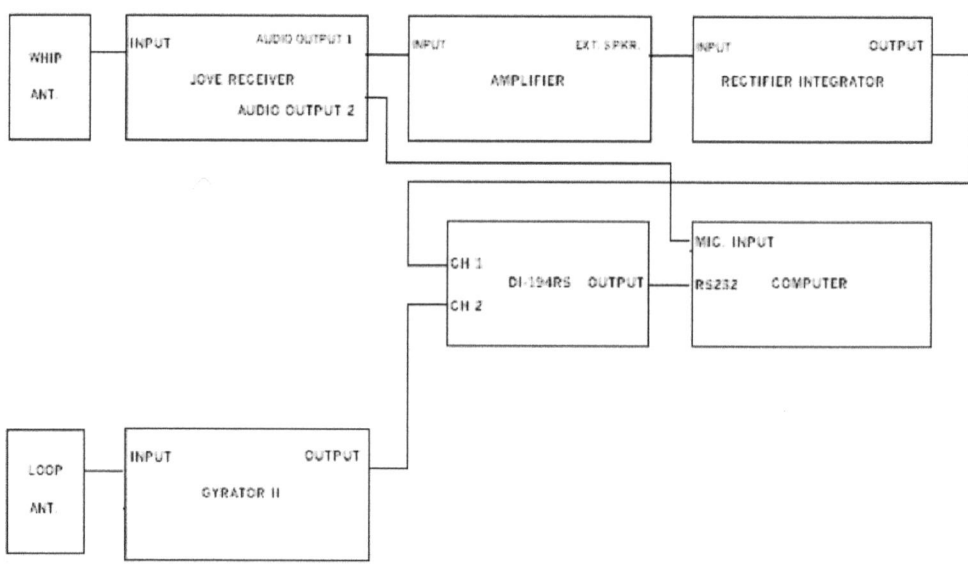

JOVE RECEIVER

All the components were mounted on a 24" by 10" white shelving board. Hook everything up per the above drawing, using 1/8" mono plug cables, except for the antenna inputs.

The JOVE receiver was assembled and calibrated per the supplied instructions. The only difference was a Radio Shack plug in power supply instead of a battery, Radio Shack PART NO. 273-1775 (http://www.radioshack.com/product/index.jsp?productId=2190632).Get the M adapter plug with the power supply, make sure the plug is installed with the tip positive.

The only difference was a Radio Shack plug in power supply was added, Radio Shack PART NO. 273-1767. Get the F adapter plug (http://www.radioshack.com/product/index.jsp?productId=2062428)with the power supply, make sure the plug is installed with the tip positive.

A 1/4" hole was drilled 5/8" down and 3/8" from the left, on the top left corner of the PC board, for an audio connector for the power supply, like the ones already used on the board.

The whip antenna for the JOVE receiver is a Radio Shack 102" steel antenna, PART NO. 21-903 (http://www.radioshack.com/product/index.jsp?productId=2102428) . A 50' lead in cable, Radio Shack PART NO. 278-971 (http://www.radioshack.com/product/index.jsp?productId=2103483), was used.

Cut one of the PI-259 plugs off and replace with an F connector, Radio Shack PART NO. 278-277 (http://www.radioshack.com/product/index.jsp?productId=2103472), so that it can be connected to the JOVE receiver. The antenna was mounted on an inverted plastic flower pot.

The amplifier is a mini amplifier (http://www.radioshack.com/product/index.jsp?productId=2062620) from Radio Shack.

RECTIFIER-INTEGRATOR

The rectifier-integrator was assembled in a Radio Shack enclosure, part no. 270-1803 (http://www.radioshack.com/product/index.jsp?productId=2062281), with audio connector input and output jacks. 1N914 diodes can be used, if you can not find the 1N60 diodes.

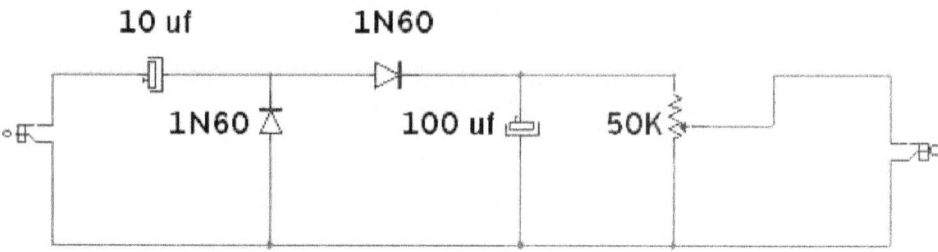

LOOP ANTENNA

The loop antenna was designed using #14 stranded wire, similar to the one Casper Hossfield describes in the April 2002 AAVSO Solar Bulletin (http://www.aavso.org/observing/programs/solar/apr02.pdf).

PARTS:

HOME DEPOT

4	1 1/4"	PVC	27 3/4"
4	1 1/4"	PVC	4 3/4"
1	1 1/4"	PVC	6"
3	1 1/4"	PVC	90 DEG. ELBOW
1	1 1/4"	PVC	TEE
4	1 1/4"	PVC	END CAP
1	1 1/4"	PVC	CROSS
1	1 1/4"	PVC	MALE ADAPTER
1	1 1/4"	GALVANIZED	FLOOR FLANGE
1	2' BY 2'		3/4' PLYWOOD
1	500'		#14 STRANDED COPPER WIRE

RADIO SHACK

    1   BOX   PART NO. 270-1801
(http://www.radioshack.com/product/index.jsp?productId=2062279)

    1   BARRIER STRIP   PART NO. 274-656
(http://www.radioshack.com/product/index.jsp?productId=2103228)

    1   SO-239 CHASSIS SOCKET   PART NO. 278-201
(http://www.radioshack.com/product/index.jsp?productId=2103295)

    1   50' CABLE   PART NO. 278-971
(http://www.radioshack.com/product/index.jsp?productId=2103483)

    1   1/8" MONO PLUG   PART NO. 278-227
(http://www.radioshack.com/product/index.jsp?productId=2103395)

LOOP ANTENNA ASSEMBLY

When connecting the PVC lightly sand both ends before applying PVC cement.

Install the four 27 3/4" 1 1/4" PVC to the cross.

Install the tee, as per the picture, to one of the 27 3/4" sections. On the bottom.

Install the three elbows, as per the picture, to the three remaining 27 3/4" sections.

Install the 4 3/4" PVC to the elbows, and the tee.

Install the 1 1/4" caps, do not cement the one on the tee at this time. This is so you can have access to the wires when pulling then to the center of the cross.

Install the 6" PVC to the bottom of the tee.

Install the male adapter to the bottom of the 6" PVC.

Install the floor flange to the center of the 2' by 2' plywood.

Drill a hole, in the 4 3/4" section connected to the tee, large enough to push the #14 wire through, next to the tee, on the underside. Also drill one next to the cap, on the underside.

Feed about 4' of the #14 wire into the hole next to the tee, this will be pulled up to the center of the cross, on the inside of the PVC.

Wrap the #14 wire around the 4 3/4" sections 21 times. Cut wire leaving about 4' to feed into the hole next to the cap.

Feed the 4' of the #14 into the hole next to the cap, on the tee 4 3/4" section.

Drill a hole large enough to pull two #14 wires in the back center of the cross.

Push the wires on the inside of the PVC up to the center of the cross, do this through the male adapter, pull the wires out the hole in the back center of the cross.

Cement the cap on the 4 3/4 tee section.

Screw the antenna assembly to the floor flange on the plywood 2' by 2'.

Mount the barrier strip on the inside bottom of the box, 3" length of the box, with super glue.

Drill a hole in the bottom of the box, above the barrier strip, to bring the two #14 wires from the cross, into the box. Pull the two #14 wires into the box, cut and hook up to the barrier strip.

Super glue the box to the cross.

Mount the SO-239 socket on the lid of the box so that it does not hit the barrier strip when installed on the box. Wire the socket to the antenna wires on the barrier strip.

For 24 khz use .0276 ufd of capacitance to tune the loop. Hook this across the antenna leads on the barrier strip.

Cut off one of the PL-259 plugs on the 50' lead in cable, and install the 1/8" mono plug. This will plug into the Gyrator II receiver.

Capacitance to tune the loop for other frequencies.

khz	ufd
21.4	.0346
24	.0276
24.8	.0257
26.1	.0233
27	.0218
30.6	.0166

GYRATOR 24 KHZ RECEIVER

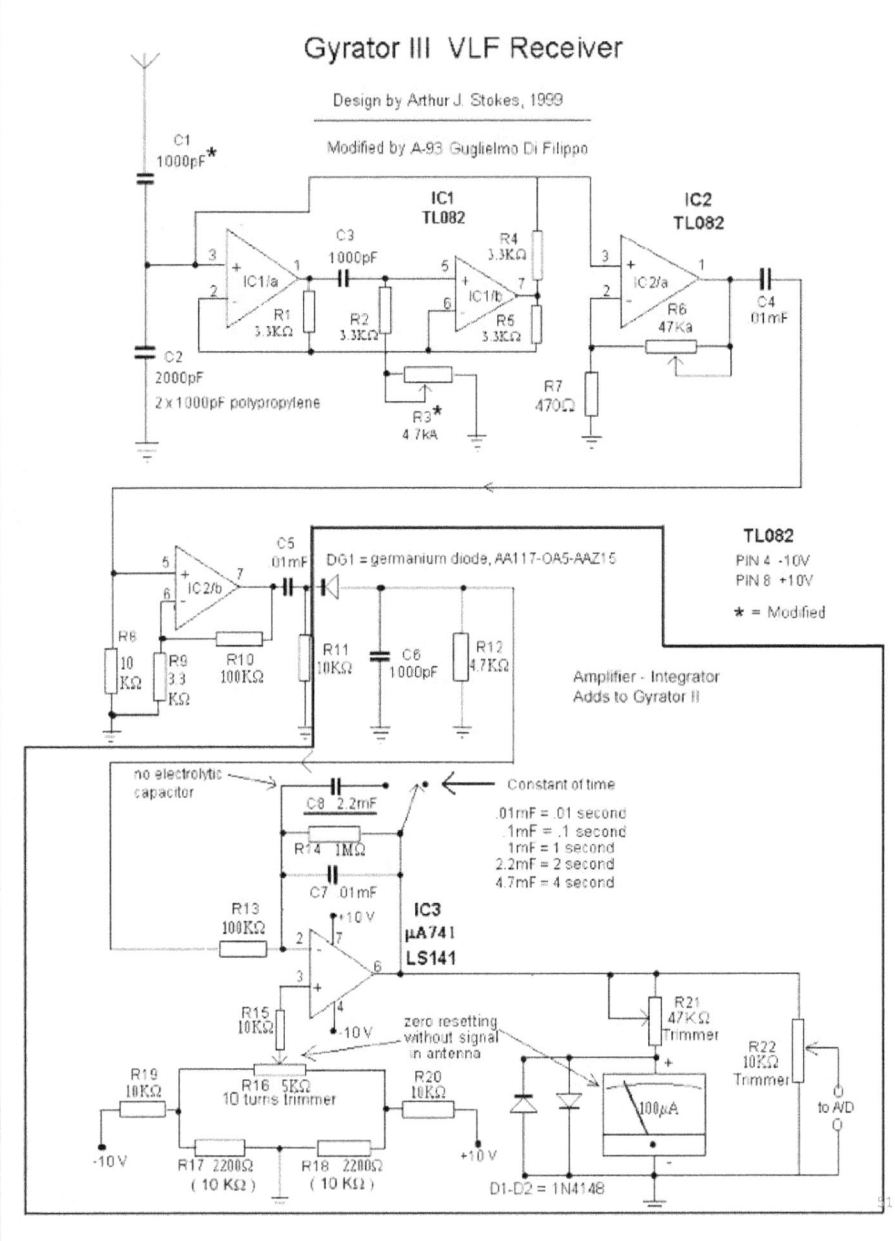

The GYRATOR III receiver was assembled per the instructions on the AAVSO web site (http://www.aavso.org/gyrator-iii-vlf-receiver). Get the pc board from FAR circuits (http://www.farcircuits.net/receiver1.htm), makes life a lot easier.

ANTENNA'S

Install the LOOP antenna and the Whip antenna outside lway from the buildings and power lines. Align the LOOP antenna with NAA in Cutler, Maine for 24 khz operation.
Run cables inside with the LOOP going to the GYRATOR III and the whip to the JOVE.

CALIBRATION AND OPERATION

Turn everything on, JOVE, amplifier, GYRATOR III, computer, and let them warm up.

Start DATAQ WinDaq 194 program.
(http://www.dataq.com/products/software/acquisition.htm)

Click View,Format Screen, 2Waveforms.

Click Edit,Sample Rate. Input 100 for sample rate.

Click OK.

On the JOVE system adjust the Rectifier-Integrator pot to full counter clockwise..

Set JOVE volume control at 10 o'clock. Tuning at 12 o'clock.

Adjust the mini amplifier volume control so there is about 3.5 volts on channel 1.

On the GYRATOR III system set the 10K tuning pot all the way clockwise, and the 50K gain pot all the way counter clockwise. Adjust the tuning pot slowly counter clockwise, until a voltage increase can be seen on channel 2.

Adjust the pot slowly back and forth until the maximum output is obtained, this should be about 5.5 volts to 6 volts. It is now tuned to 24 khz. Turn the 50K gain pot clockwise until there is about 3.5 volts on channel 2.

Click Scaling, Limits. Input channel 1 top limit at 6, and bottom limit at 0.

Click next. Input channel 2 top limit at 6, and bottom limit at 0.

Click OK.

Click Options, scroll mode.

Click edit, Sample Rate. Input 1 for sample rate.

Click OK.

Click File, Save Default Setup.

Click File, Record. Input file name.

Click Open.

The file size can be adjusted for a longer or shorter run at this time.

Click OK.

ENJOY

Solar Observatory

Due to the fact that I am follicle challenged, bald that is, I needed a solar observatory for my Coronado MaxScope 40 (http://www.coronadofilters.com/), more about the scope later. While wandering around the Bass Pro Shop (http://www.basspro-shops.com/servlet/catalog.OnlineShopping?hvarClearAffiliate=yes), I saw this out-house hunting blind that would make a small portable solar observatory for my MaxScope. The observatory will work with any small refractor for day or nighttime observing. I had a mechanic's table from Wal Mart, the table top was adjusted so my monitor would slide into one of the windows on the blind. With the blind staked out you have enough room for the scope, a chair, and eyepiece case. Maybe next summer I will get a small air conditioner, and put in the other window. The good life.

Donated Sun Guns Gallery

Sun Gun Goes To Washington
Smithsonian Institute

I donated a Sun Gun to the Smithsonian National Air and Space Museum, Public Observatory Project (POP). (http://www.nasm.si.edu/exhibitions/popobservatory.cfm) Hopefully it will encourage young minds to pursue science.

POP Youtube video.
(http://www.youtube.com/watch?v=Bm1kqAhf-88/)

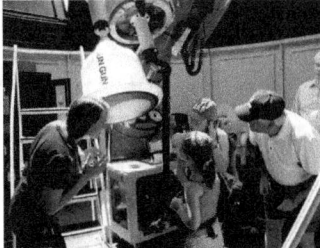

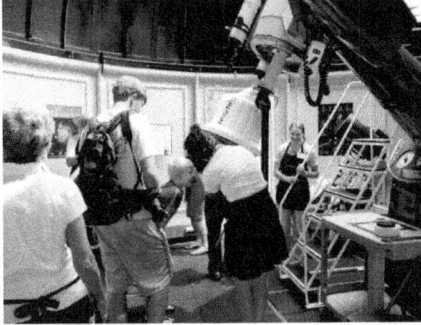

South Carolina State Museum

I donated a Sun Gun to the South Carolina State Museum.
They have a large collection of Alvan Clark telescopes, a must see.

Discovery Museum and Planetarium Bridgeport, CT

I donated a Sun Gun to the Discovery Museum and Planetarium Bridgeport Ct. Stop by and see it when in the area.

Fernbank Museum

I donated a Sun Gun to The Fernbank Museum.
Stop by and see it when in town.

The Franklin Institute

I donated a Sun Gun to The Franklin Institute.
Stop by and see it when in town.

Sun Gun Goes to Saint Louis Science Center

I donated a Sun Gun to the Saint Louis Science Center.
Stop by and see it when in town.

Exploratorium

I donated a Sun Gun to the Exploratorium (http://www.exploratorium.edu/). When in San Francisco go to the Exploratorium the best scientific museum you will ever visit.

Sun Gun Goes to Adler Planetarium

I donated a Sun Gun to the Adler Planetarium, Doane Observatory (http://www.adlerplanetarium.org/). When in Chicago this place is a must see.

Sun Gun Goes to Tellus Museum

I donated a Sun Gun to the Tellus Museum, (http://www.weinmanmuseum.org/) in north Georgia. It's a fantastic facility that is a great place to encourage your interest in science. Besides which their five cheese grilled sandwiches are out of this world!

The Environment

In addition to his interest in astronomy, my father was passionate about the environment and alarmed by the harm being done to it as evidenced by this poster he made.

Here are links to some of his favorite causes:

GIVE A RIBBIT, HELP THE FROGS

http://www.atlantabotanicalgarden.org/conservation/amphibian-research
http://www.savethefrogs.com
http://amphibianrescue.org
http://nationalzoo.si.edu/SCBI/SpeciesSurvival/AmphibianConservation/
http://www.amphibianark.org

GIVE A BUZZ, HELP THE BEES

http://www.xerces.org/bumblebees/
http://bumblebeeconservation.org
http://extension.uga.edu/publications/detail.cfm?number=B1164
http://beeconservationnigeria.com
http://www.conservationevidence.com
http://www.naturalcuresnotmedicine.com/2014/06/make-bee-waterer-help-hydrate-pollinators.html

www.ingramcontent.com/pod-product-compliance
Lightning Source LLC
Chambersburg PA
CBHW061219180526
45170CB00003B/1066